Astrophysics

In the Form Of

A Story

UNDERSTANDING OUR UNIVERSE IN AN
INTERESTING WAY

HITEN SHELAR

Copyright © Hiten Shelar 2024

All Rights Reserved.

This book has been published by IMAGINE SPACETIME (www.imaginespacetime.com) with all reasonable efforts taken to make the material error-free by the publisher. No part of this book shall be used, or reproduced in any manner whatsoever without written permission from the author, except in the case of brief quotations embodied in critical articles and reviews.

The Author of this book is solely responsible and liable for its content including but not limited to the views, representations, descriptions, statements, information, opinions and references ["Content"]. The Content of this book shall not constitute or be construed or deemed to reflect the opinion or expression of the Publisher or Editor. Neither the Publisher nor Editor endorse or approve the Content of this book or guarantee the reliability, accuracy or completeness of the Content published herein and do not make any representations or warranties of any kind, express or implied, including but not limited to the implied warranties of merchantability, fitness for a particular purpose. The Publisher and Editor shall not be liable whatsoever for any errors, omissions, whether such errors or omissions result from negligence, accident, or any other cause or claims for loss or damages of any kind, including without limitation, indirect or consequential loss or damage arising out of use, inability to use, or about the reliability, accuracy or sufficiency of the information contained in this book..

Published by IMAGINE SPACETIME (www.imaginespacetime.com)

CONTENTS

Preface

About The Book

The Story and Purpose Behind the Creation of "Astrophysics In The Form Of A Story".

1) What Is Time And Why Can't We See It?	15
2) What Is Gravity And How Is It Related To Time?	19
3) Why Time Does Not Run For Light?	25
4) Bizzare States Of Matter, With Its Startling Nature!	32
5) The Mysterious Regions In The Universe Where Reality Breaks!	40
6) Black Hole As A Portal To Parallel Universe!	44
7) The Waves Of Gravity!	52

8) Mass To Energy Transformation Without Nuclear Fusion! — 56

9) The Lens Made Out Of Gravity! — 61

10) The Mysterious Substance That Is More Abundant Than Normal Matter! — 67

11) The Birth Of Time, The Birth Of Our Universe, And The Big Bang! — 71

12) How Far Is The Edge Of The Universe And Where Is Its Center? — 75

13) Is Reality Real? The Bizarre World Of Quantum Physics. — 79

14) Why This Universe Is The Way It Is? — 83

15) The Laws Followed By Every System Present In The Universe. — 88

16) Universe Breaking Its Own Law Of Physics! — 92

17) How Does A Gravity Tractor Move An Asteroid From Its Orbit? — 96

18) Time Travel And The Spacetime Inside A Black Hole. — 102

19) The Strange World Witnessed By The Light And Gravitons. — 108

20) After the event horizon of a black hole. 111

21) Does God exist? 114

22) Are we alone in this universe? 116

23) The Future Of The Observers Creating This Universe. 121

PREFACE

Hello! I would like to start with a bunch of facts that will make you curious. Do you know how much our earth weighs? Yes, have you ever even thought about that? The mass of our planet Earth is around 6 followed by twenty-four zeros! This is just a huge number if you compare it to the mass of a human. I mean have you ever questioned to what extent should something be made up of something? For example, let's take your shirt, let's zoom into it and what you will find is a set of molecules, zooming further into the molecules you will find sets of atoms forming that molecule. Now, make this story go on ahead. My main question now is where does this all end?. Again, have you ever thought about how large can something be or if does universe itself has a boundary like we have for different territories on our planet? Did you know that the stars we see in the night sky are the same thing as our sun? And almost all of them are larger and more massive than our sun. And also they have planets like our star system has. But don't confuse these stars with existing planets in the sky. I mean planets look as same as stars in the night sky "a dot". The only way you differentiate them without using a telescope is by looking at them to check whether they twinkle. If they don't twinkle they are planets and if they do, they are stars! Do you ever hear about the unreal interstellar objects? I mean have you ever heard of neutron stars, pulsars, magnetars, black holes, and white dwarfs? These things have an unreal nature because they are giant structures not made of normal atoms. Do you ever have a question about who made this universe in the first place? Does god exist? This book composes many other questions like these and the answers to them, that will blow your mind.

This book is composed of all the astonishing facts like these for every human who asks questions about our universe while looking at the dazzling night sky. So let's dive into it.

ABOUT THE BOOK

You are a beginner to astrophysics or in some other profession than Astrophysics. You always realize that you have unanswered queries about our universe whenever you gaze at the stars. Again, if you want to get the answers to your questions in a hurry, you are a beginner, you are not a physicist or you don't like maths or I but want to know and get your questions solved about our universe. And more importantly, if you like to learn science in the form of a story. Then this book is for you. This book just not will only answer your unanswered questions about stars, galaxies, and planets, but also take you to the regions of the universe where the real universe seems unreal. This book talks about everything, every human should at least know about our universe. This is not a standard academic astrophysics textbook, it's a ticket for everyone to satisfy their curiosity about knowing our universe in an interesting way. The world always needed this.

STORY AND PURPOSE BEHIND THE CREATION OF

"Astrophysics In the Form Of a Story"

In the early days of January 2023, I embarked on an incredible journey. I set out to write a book, not just any book, but the proud recipient of the International Best New Cosmology Books Award for 2023—"Comology for a Newbie."

Hailing from the vibrant city of Mumbai, I felt the urge to reconnect with the universe profoundly. So, I packed my bags and headed to my peaceful village, surrounded by nature and an unobstructed view of the starry night sky. It was there that I found the perfect canvas to put my thoughts into words.

Writing a book on astrophysics has always been my passion. I believe in making complex concepts simple and accessible for everyone. As I delved into the process of writing, I had a delightful surprise. I decided to chat with my grandpa about astrophysics, half-expecting him to be disinterested. To my astonishment, his eyes sparkled with curiosity, and he bombarded me with questions about the universe.

But it didn't stop with my grandpa. Through my books and social media, I've had the privilege of reaching hundreds of thousands of people and sharing the enchanting world of astrophysics with them. There's something inherently captivating about astrophysics—it's a subject that transcends age, background, and boundaries.

We've all gazed at the night sky, marveled at the shooting stars, and felt the wonder of Saturn's beautiful ring system. Yet, during our school years, astrophysics was often given limited exposure or omitted altogether. This book, "Astrophysics in The Form Of a Story" is my attempt to change that.

Nowadays, astrophysics is looked more commonly as a tough academic major instead of a curiosity. Throughout my experience as a science communicator, it's been very clear that people love to learn such complex concepts when they are explained in the form of a story. For example, one may give a black hole a picture of a sci-fi portal to explain the concept in the form of a story. Every day my little sister and meet, and we enjoy this conversation about astrophysics. And she really loves to hear these concepts in the form of a story. And I could explain it to her effortlessly. So no matter if you are kid, a teen, an adult or an old man, we are gonna love to learn about our universe, explained in the form of a story. It's In the form of a story so it's much of a popular science than an academic one. As a science communicator, I have been making science easier for people through my bestselling books Astrophysics for non-mathematicians,

Cosmology For A Newbie, etc. Creating "Astrophysics In The Form Of a Story" is another action of mine to make science easier for people.

Everything we call real is made of things that cannot be regarded as real.

- **Neils Bohr**

ONE

WHAT IS "TIME" AND WHY CAN'T WE SEE IT?

The time which we use and base our schedule on, which can be a clock at your home or wristwatch is invented by humans. The time on our clocks is not what time really is. Clocks are just made to keep a record of running time. This running time is an actual or real-time I am talking about. This "time" was there even before any human ever existed. The creation of this time dates back to the birth of this universe. And of course back then we had no clocks and calendars. So the very basic definition of the "real-time" I can give you is that, "everything in our universe ages". Now you invent home clocks or wristwatches or not, everything is still going to age and so time is always going to run. The clocks were just invented to make a note of running real time so that daily activities be done during the availability of sunlight and sunset or any other example one might relate.

Sir Issac Newton, It is OK if you don't know precisely what his work and contribution to physics is. But one thing I can tell you for sure is that we all know him as one of the greatest scientists of all time. Now, Newton said that real time runs the same for everyone in the universe, no matter where you are. Now you will tell me, of course, that it is so obvious. Let me tell you a surprising fact for this chapter. The fact that you may not digest if you don't firmly believe in me. The fact that time or real is not absolute! It runs differently based on a few parameters. Let me give you an example of it. If a person stays on

Earth and his twin goes on planets like Jupiter. Then few years after they reach home, their age would no longer match! I mean there is nothing special in Jupiter in this case. Just bear with me for some time. One more example I can give you is that if one twin takes a round trip somewhere in the universe and back to Earth and so that he travels at a particular range of speed (near to the speed of light) makes that twin younger than one who stays at Earth.

The above few sentences might have forced you to review again and again what's going on if you couldn't digest it. But this is just one of the fundamental facts of this bizarre universe. Now when I took the example of Jupiter, there is no presumably special power in Jupiter. It was just an example. This happens with every massive body including our earth (even if not that influential). In the next chapter, we look at why this happens and believe me it's going to open your eyes and make you stay awake the whole night thinking about our universe and its bizarre nature. In the second example, I mentioned the speed of light. This is because the faster you move in your rocket or spaceship the slower the time will run for you. We will soon discuss this in our later chapter. The relativity in time which I discussed here was brought into the vision first by no one but one of the greatest scientists of all time, Albert Einstein in his theory of relativity.

WHY CAN'T WE SEE TIME?

The world we look at around us has three dimensions. Everything we look at has a length, breadth, and height. The reason behind we don't talk about more than three dimensions is that because our mind just cant think about more than 3 dimensions. Albert Einstein's theory of relativity says time is four-dimensional. And that's why we can't see time.

Before we move to the next chapter, let me drop one more bombshell for this one. The fact that "time does not run for light!". Shocked!? But for now, I request no further discussions on this, just bear with me as we finish reading this book to know how is this possible.

TWO

WHAT IS GRAVITY AND HOW IS IT RELATED TO TIME?

There is one thing common between time and gravity, that both of these things can't be seen although we definitely know that they exist. In the previous chapter, I explained why we can't see time; because our mind can't think more than three dimensions, and time is the fourth dimension, we can't see it. But what about gravity? Okay, now I have to explain what is gravity to the reader. The one thing that will surprise you in this chapter and make you stay awake at night is knowing what gravity alone is, so just bear with me.

Now I want a reader to look around himself in all the directions possible. This is our 3-dimensional reality, which we can see and think about. We call it space (Let's call what lies beyond Earth "the outer space" so that we don't confuse ourselves as we discuss.). But it's not the entire thing that our universe is. The space around us is what we can see and think, but what about something we can feel? Of course, that thing which we can't see due to limitations of mind but can feel is the "time". And this "time" is also a dimension. So in totality, we have 4-dimensions. So we live in a 4-dimensional reality which we call the universe. We also call it **"the spacetime"**.

Now to explain what gravity is, imagine this 4-dimensional reality or spacetime as a 2-dimensional sheet of rubber or cloth. Where, instead of three space dimensions we take one space and another time. Now imagine holding this cloth tightly and putting a heavy ball on it and now you will see some curvature in the sheet of rubber or cloth. Now throw a few marbles over this cloth in arbitrary directions. And you will see them taking orbits around a ball in the center. Compare this with the earth and all the planets taking an orbit around the sun and also the earth and moon system. Well, this is what gravity is!

(Fig. 2.1) The Grid lines shown in the figure above are the fabric of spacetime

The rubber sheet in the example we have taken is nothing but 2-dimensional spacetime. In the real universe, this happens with 4-dimensional spacetime, which in fact our real universe is. So is gravity a curvature in reality? The answer is yes! This curvature is but what we can't see, since we all are in the same curvature of spacetime. This can be compared with travelling it a train, where your friend and you may not find yourself moving inside the train, but your other friend outside the train will do. Although we can spot these curvatures by looking at the far objects that don't share spacetime curvatures with us.

(Figure 2.2) Light is influenced by the spacetime curvature around the astronomical body creating it, showing its properties in the form of a distorted image. The reason behind showing this image in this chapter is to give a real-world idea of spacetime curvature to the reader.

So everything that has mass, curves the spacetime around it including me, you, and everything. Except for the fact that the spacetime curvature we possess is very far away from the matter of consideration.

Spacetime curvature possessed by a human or daily object ≈ 0

The spacetime curvature created by the earth and sun is huge enough to make us feel gravity the way it is.

WHY IS TIME NECESSARY FOR GRAVITY?

In the example, we have taken all the thing that was dragging marbles to the ball in the center Is the weight of the ball. The weight of the ball is a measure of how strongly it is pulled by gravity which is a measure of how massive a ball is. So, in our illustration of gravity, we just were explaining gravity with the help of gravity. **Although this analogy is fair enough to explain what gravity is**. In the original picture, we replace the gravity of the earth which we took in our example by the "flowing time"

. And as we know time always flows, and gravity never halts itself no matter what! The intriguing fact of this chapter is that Gravity can be sent in the form of a wave! This gravity wave can be sent wherever one wants in the universe. The reader might say "What!?". But don't worry I am going to discuss all of these facts throughout this book. In the next chapter let's discuss why time does not run for light.

THREE

WHY "TIME" DOES NOT RUN FOR LIGHT?

The "light", the samething with the help of which the reader is able to read this book is an electromagnetic wave. It sometimes also behaves like particles. Come on, just wrap this up! Why? Because there is nothing special in the light, except the speed at which it is traveling!

We know that our universe is unthinkably huge in dimensions. And one can travel wherever one wants in this universe. But did you know that there exists a limit on how fast can anything travel in this universe? I am not talking about the limits of our rocket engines or propellers. I am talking about the true fundamental limits imposed by the universe. Oh! Did you even know any such fundamental limits ever existed? These fundamental limits are in fact the reason physics works the way it does! In this chapter we will discuss the fundamental limit of speed one can travel at

The fundamental speed limit or universal speed limit we are talking about is a speed called "the speed of causality" denoted by "c" Its value is around 3,00,000 km/s. This speed is the fastest speed can anything travel in this universe. The reader will say, come on I will press my accelerator at 3,00,000 km/s and I will simply break this limit. And nowhere comes the bizarre nature of the universe.

We are here sitting in front of a book, on our couch, in a state, not the best to see the nature of the bizarre universe, many people even can't move themselves to check what's happening in the night sky. I still remember a day when I was on the terrace of my house in Mumbai in January 2021. That day due to some unique weather behavior patterns, Mumbai city while being one of the most polluted cities in India was witnessing one of the clearest skies I ever witnessed. That day, even we could see the Mercuryset (setting mercury on the horizon) which is impossible to see in Mumbai on other days. That day I called my friend to see Mercury and he refused just because he was feasting at his home and he missed the chance. Since then, till the day I am writing this book (2023) I haven't witnessed a sky that clear in the metropolitan city of Mumbai. Again, these astronomy observations are just not observations, but instead, a thing of beauty which as John Keats said is joy forever. Sitting on our couch we are so prone to the world the world around us that we start to think of it as the entire universe, which in reality is far more bizarre than one might think. Let's come back to our chapter.

The speed of causality is not as easy as pushing down the accelerator. Firstly. reaching even near to such a speed is a great deal. Now let me tell you how. I hope the reader has not forgotten the concept of spacetime I explained in the last chapter. If forgotten, then just take a look at the last chapter where the word "spacetime" is highlighted boldly.

The speed of the fastest rocket we have ever launched is not even near to the speed of causality. But, it is very common in the case of atoms and subatomic particles to move at near to speed of causality. So before we move to why time does not run for light. Let me give you some facts of what happens when one moves near to speed of causality speed. So when you reach a speed to is considered comparable to the speed of causality following things happen:

- **Length contraction**: A spacecraft traveling at a speed comparable to the speed of causality (3,00,000 km/s) will see objects in the surrounding universe having their length shrunk. For, example if looked at a meter-long rod, depending on what speed one is exactly traveling at, it would appear shorter than a meter. The same thing will people even see outside the spacecraft when they look at the spacecraft. Let's call these people "external observers". They will see the length of the spacecraft shrunk in size than what you might know as a traveler.

- **Time Dilation**: As one nears the speed of causality, the ticking of time (I mean real time) for him/her slows down. This gives rise to things like the twin paradox, where one twin move near to speed of causality, and the other stays at rest throughout the roundtrip, at the end of the roundtrip they have different age! As we go nearer to the speed of causality, slows further down, and perfectly at the speed of causality the time does not tick at all.

- **Increment of mass**: As one travels and reaches a speed comparable to the speed of light, his mass increases when seen by external observers. This is again the reason why normal-day objects can't move at a speed comparable to the speed of light. Since the mass keeps on increasing, more is energy needed to keep pushing the spacecraft. At very near to the speed of causality, the mass becomes infinite even in the case of subatomic particles and now the only possible particles that can possibly travel at the speed of causality are th the ones that have zero mass.

I can end this chapter here now itself (although I am not doing it) by simply saying that "light travels just at the speed of causality". Okay, so to elaborate, light travels at a speed of causality which is the fastest speed possible in our universe. And as we discussed when you move at the speed of causality, time does not run for you. Since light travels at the speed of causality, time does not run for light. Now, one might say that light takes around 8 minutes to reach us from the sun. But let me tell you, if you ask the same photon, it will simply say "No time elapsed!". Before we move to the next chapter, I wanted to drop one more intriguing fact here in this chapter. The fact that we can travel faster than light without actually traveling and reach our destination by not yet built technology of warp drive. One more intriguing fact? If you want to experience what light can experience, you can actually visit a place in the universe where the same thing can happen to you. It is a place, if visited, time will stop for you. Astounding right? Let's dive further into the ocean of these facts now!

FOUR

BIZZARE STATES OF MATTER, WITH THEIR STARTLING NATURE!

I want a reader to take a look at the surroundings, and I bet on your couch everything you might be seeing is made of something called molecules, which are made of atoms. These atoms can be hydrogen, carbon, oxygen, nitrogen, or any existing atom. These are normal atoms that kind of have the same type of structure and work on the same principle as normal atoms do. But can you imagine thinking of matter, a spoonful of which might weigh as heavy as Mount Everest? I think but this shouldn't be called "matter". Since it shows no similarity in behavior in any kind of sense with matter. And again what I am talking about is real. Let's further know what this type of matter really is.

We all know or I request readers to know that stars are not a ball of fire. They can instead be seen as a nuclear reactor where the nuclear energy harnessed is used to maintain the star's own gravity. Yes, the more massive and dense something is, the stronger and more the spacetime curvature it creates. If you again forgot what spacetime is, please check it on page no.18. And the more the spacetime curvature, the stronger the gravity. Remember, the entire solar system is based on the gravity of the sun. So what maintains the outer layers of stars from collapsing inwards due to gravity is the same nuclear energy as just discussed above. And the excess energy gets delivered to Earth in the form of light. Which outshines the entire day, creates weather patterns, provides food for plants, and creates life. The primary energy Earth receives is from the sun and then the cycle of the ecosystem goes on. This all was just to explain that the sun is not a ball of fire. Also, for fire one needs oxygen for combustion. Again, the sun is not a ball of fire! Let us move further to our mystical matter which seems to have a totally different physics.

The tug of war between gravity and nuclear energy of the sun or any star in the universe, in fact, is how a system that we call a "star" works. But a star always has a limited amount of fuel, once exhausted, gravity wins the tug of war. When this happens the immense gravity of the star shows its effects, crushing every point of a star so much that the matter which the star is made of, itself gets shattered. When a star as massive as our sun goes through this process, in the end when gravity crushes everything, It forms "A whiter dwarf". This entity of white dwarf is not made of anything under the senses of nonphysicists. The white dwarf star is made of a totally different state of matter.

Normally an atom is composed of an electron gas around a nucleus. But when this atoms are crushed altogether by the immense gravity of a star, it forms a white dwarf. This white dwarf is made of nucleons and especially the extremely pressurized electron gas. And according to physics, there is a limit on how much you can compress this electron gas, this limit maintains the surface of a white dwarf from inward collapse. Again, if a star is more massive than our sun, there is still a tug of war-between the matter of a white dwarf and its own gravity. And provided we have a star enough massive, so that its gravity again wins the race, we form "A Neutron Star". Again, it is made of a totally different state of matter.

An atom is made of electron gas orbiting nucleons, nucleons are made of protons and neutrons. When a proton combines with an electron it forms a neutron along with a neutrino. Neutrinos are ghost-like particles that interact so weakly that they pass literally like a ghost from anything they encounter. Let me tell you right now, there are trillions of neutrino particles passing by you every single second, not interacting with you due to their weak interaction. These neutrinos are created by the sun, which creates them in the process of nuclear fusion. Neutrino has a negligible chance to interact via the weak force.

When a star crushes itself into a neutron star, by collapsing the electron gas of a white dwarf with available protons in the nucleons, it produces an enormous amount of neutrinos. These neutrinos but here remain interacted with the inward layer of stars in this case, due to the high density of stars (increasing the chances of neutrino interaction). And as these neutrinos flux out of the star, they carry the upper layers of stars along with them. Shredding these layers in one of the most violent events of the universe " A Supernovae"!

These events if seen from Earth at a safe distance can look incredibly beautiful, when move yourself from your couch to go on the terrace and look up at the cosmic window available to every living creature on our planet "Our sky". These events can brighten the entire sky even at night. These beautiful events have already been witnessed by humanity. The one which is SN 1054.

Anyways, coming back to our chapter, this all happens when a star massive enough forms a neutron star. The matter of these neutron stars is difficult to imagine but can be thought of as made mainly of neutron gas. Which again has a limit on how much it can be compressed, a process which again is responsible for the stabilization of neutron stars. It was the state of matter of neutron stars that I was talking about, a spoonful of which weighs as heavy as Mount Everest. Such states of matter are incredibly hard to visualize, unthinkable to think how a system mostly made of neutrons would look like. The state of matter of the white dwarf if it gets to Earth, will not look less mysterious than a traditional science fiction mysterious material with the nature of somewhat showing different powers, unlike regular matter. We are yet to discuss the most intriguing fact of this chapter.

This is how bizarre the nature of our universe is when you explore things in the night sky. Some really interested people in the world, now are exploring radio astronomy. This gives you a totally different lens to look at the universe, looking at things that can't be seen in visible light. Some are now even using gravitational wave detectors. Some are using the arrays of pulsars (Pulsars are rapidly rotating neutron stars emitting light from their poles) to look at the period after when our universe was just born. And the further we dive, bizarre the things are gonna get in Astronomy. Returning back to our chapter.

Wait! What If the star is massive enough to collapse the neutron star further? I mean, what would we end up creating? This will be discussed in the next chapter. But here I can't tell what the true nature of this thing is, if we could get it to the earth. Which in fact is also an intriguing fact for this chapter. It would look like a mysterious portal! Portal!? Portal to what!? Let's discuss that in the subsequent chapter.

FIVE

THE MYSTERIOUS REGIONS IN THE UNIVERSE WHERE REALITY BREAKS!

Do you know there are few portal-like things available in the universe? If we go near it, it will look like a pitch-black spherical ball made out of no material particle. And I am calling it portal-like because it swallows whatever it encounters in its path and the object swallowed has no way to communicate with the outside universe. Also, it has mass and gravity. It not only swallows whatever comes in its path, but it forcibly attracts everything to come into its path to swallow it by its gravity. The portal I am talking about has the strongest gravitational influence than any other object in the universe. I am talking about "A Black Hole"!

I can't call it portal because then I have to answer the question "Portal to what?". Because as per our scientific evidence, all we can see is the pitch-black portal-like thing (since it looks like a portal). Black Holes have gravity so strong enough that even light cannot escape it. Light curls around with spacetime in its orbit, entering the black hole at the end.

But what this Black Hole really is and what creates it? Ok, in the preceding chapter, we discussed White dwarfs and Neutron stars. But what if a star is massive enough that in a tug-of-war between a neutron star and the collapsing gravity of a star, gravity again wins? What we are left with now? Until now we just have been discussing the collapse of a star. But as it gets denser, we do not even care about its influence on the fabric of spacetime (which our universe is made of). Mass curves the spacetime. The denser the mass gets, the stronger the spacetime it generates around itself. Throughout the process of transformation from star to white dwarf to neutron star, there has been a tremendous increment in the density of a system. For a star like our sun (one solar mass) which has a radius of 6,96,340 km, the radius of the white dwarf formed from the sun would be between 6000-12000 kms wide. For a larger star with a mass 4 solar masses and an even larger radius than our sun, the neutron star formed would be just 10-15 kilometers in radius. This density shift is immense! Imagine how strongly the spacetime curved itself incase of such a system.

But there is a limit on how much tension we can put over spacetime by mass-energy density. So, once the limit is crossed, we end up creating an infinite space-time curvature which we call "a singularity". And once that happens the time in spacetime loses its meaning, it simply breaks down. Every singularity created always has a three-dimensional black boundary around it. This black boundary is the portal we are talking about. It's called "an event horizon". The event horizon together with singularity is what we call "A black hole" It is amazing to know that we are talking about spacetime, which our real-world universe is, and we literally are discussing it like a sheet of paper!

When we launch our rockets into space we have to make sure that the rocket travels at a specific speed of 11.2 km/s, if traveled slower than this speed, rockets can't go in space. This value of this velocity is as same as the maximum speed possible in our universe (speed of causality: 3,00,000). And that's why nothing that touches the event horizon of a black hole can escape it.

An intriguing fact for this chapter is that time stops for you at the event horizon of a black hole! Like time stops for light.

SIX

BLACK HOLE AS A PORTAL TO "PARALLEL UNIVERSE"!

A Black hole is composed of an event horizon and a singularity. The physics of singularity is yet not understood. There are a lot of theories describing it, none of which is proven true. But every scientist as a human believes firmly in one of the theories. I do the same.

Singularity is the point where time loses its meaning, and so there is no time dimension at that point. So what is out there is what I shall discuss later in this book. So as I discussed, at the event horizon of a black hole, time completely stops elapsing. Let's now go with the intriguing fact of this chapter! Once you cross the event horizon of a black hole space becomes time and time becomes space! So, you form time-space (opposite of spacetime) inside a black hole. The amount of time of spacetime flown inside a black hole then projects itself in the form of three-dimensional reality. So, you form time-space (opposite of spacetime) inside a black hole. In such a time-space reality one can even look in the past! Not going into this in detail in this book, since this is an easy-level book.

What if I tell, you that like our universe, there can be many universes in a domain called multiverse? Yes! Our universe may be among many universes in the multiverse. The idea of multiverse dates back to ancient Indian and Chinese literature. Now imagine a random universe in a multiverse, a universe that even has its own black holes. Now as shown in the figure below, imagine the singularity of a black hole in our universe joins itself with the singularity of the black hole in another universe. And now by joining two singularities of these black holes of two individual universes, we kind of created a tunnel to go from universe to universe. Isn't that fascinating? This tunnel is called "a wormhole".

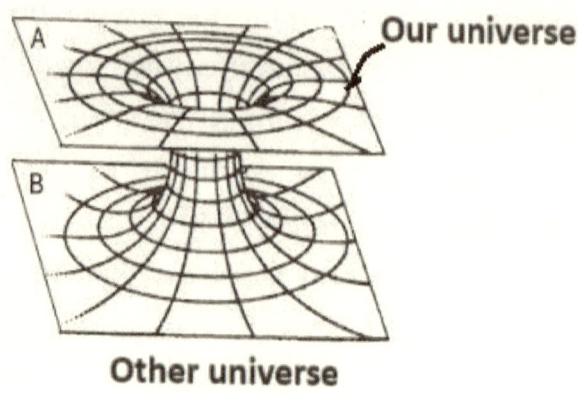

But to make this interstellar travel safe we have to make sure that the laws of physics in the universe we are travelling to have to be as same as ours with no room for error. Since anything that exists in our universe as matter-energy should have its existence sustained by the laws of physics there. For example, In other universes, time can even be running backward, the value of pie can be different, or you have 3 dimensions of space and two of time. It can be literally anything! So to make a safe inter-universe voyage we have to make sure that the laws of physics in other universes are as same as ours. This chapter has provided a lot of intriguing facts to the reader. But let's go with another one. So, if I ask you to destroy for example a piece of paper as an entity. The process you might do can be tearing the paper into pieces, burning it out, or anything. But you still don't destroy it at the end since what you are left with now is the particles of that paper that were part of that paper. But, now if you want to destroy the entity of paper leaving no clue that it was a paper ever before, the one thing one can do is by throwing it inside a black hole. And all the information of paper that will stay constant in the universe would be the mass of the paper and the charge it had, else everything becomes part of the black hole itself. This happens with everything that enters a black hole and becomes a part of it. *Black holes probably are the only macroscopic things that are huger than enough to spot and are far beyond our sense of reality.*

Black holes have a wide range of mass, from microscopic to thousands of times of size of our solar system. Smaller black holes are more violent than larger ones. Violent black holes in the sense that anything that goes closer to these black holes gets stretched like a spaghetti due to the special effect of gravity, which we call tidal gravity. Tidal gravity is the reason behind the formation of Saturn's rings, it is again a reason why the moon just shows one side of it to the Earth while rotating around its axis simultaneously. Again, the same force by the Moon on Earth causes the sea tides on our planet. The larger the black holes are, the less intense is their tidal gravity. So it is always safer to voyage through larger black holes than smaller ones. We have taken two images of a black hole with our event horizon telescope, one which resides at the center of the M87 galaxy and the other at the center of our own Milky Way. See the image below.

The black hole at the center of the M87 galaxy.
(Captured by the event horizon telescope)

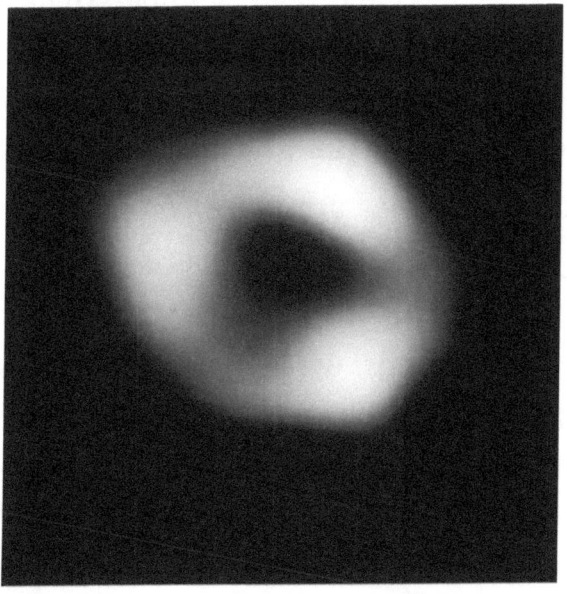

The black hole at the center of our Milky Way galaxy. (Captured by the event horizon telescope)

These two images taken are one of the greatest leaps of mankind. I would suggest readers watch "Interstellar Movie", it's all about black holes and wormholes. It is one of the greatest movies ever made by man. The black hole in the movie is created so scientifically that it is the same as what a black hole would look like in the real universe. One should understand that black holes are not the real objects at all! They form after the death of a star, which is real, but black holes themselves are a break in reality. For now, our all equations of physics fail at the singularity. In the future, we might understand these singularities better. In reality, black holes are more complicated since they are just not holes in space but a hole in spacetime. Since our mind can't think of more than 3-dimensions, we just see a 3-dimensional spatial part of a black hole. When we see the effect of a black hole on time, under some conditions it is seen to govern the time of everything that comes enough close to it. So when we say a complete governance of time by a black hole, the black hole now becomes a moment in time of a voyager instead of a location in space. And so, the voyager now has no option to avoid it in any way. In my Astrophysics for non-mathematicians, this voyage has been discussed in detail. In the same book the singularities of black hole are also discussed, and what other theories has to say about it. Dropping another intriguing fact for this chapter, a fact that like light, there is one more thing that travels at the speed of causality or the maximum speed possible in our universe. I am talking about "The waves of gravity!". Now let's discuss it further!

SEVEN

THE WAVES OF GRAVITY!

Every mass possesses spacetime curvature and spacetime curvature is gravity. We all are under the spacetime curvature of Earth, I mean its gravity. But the gravity we are influenced under is of constant magnitude. I mean the value of gravity stays the same for our entire body here on earth. Now imagine the value of gravity on earth is different for a person's head and toes, in such conditions, a person's toes would be pulled more than his/her head. And so a person will tend to stretch due to gravity. Although the earth's gravity is much much smoother, there is no difference in the earth's gravity for the two ends of Burj Khalifa (Tallest building as of 2023). Even when you travel 40,000 feet above the ground on an airplane, the gravity is 99.9% its value on the ground. But when you talk about bodies as wide and massive as the moon, there is a difference in the gravity on one part and another. Again, a reason why the moon always faces us showing its one side and not the opposite one. But this force is what we are talking about is tidal gravity. For the effect of tidal force to occur, or to be apparent there are a number of factors. Factors like the mass-energy density of the influencer, how far the influenced body and how wide it is. This chapter is on "The waves of gravity". I discussed tidal gravity, just because the waves of gravity influence anything like tidal gravity. Wait, but what these waves of gravity are? Let's discuss that further.

Throughout this book when we talked about gravity we have taken spacetime into account. This is because gravity itself is the spacetime curvature. So when we talk about gravitational waves, they are just waves through spacetime. Compress the 4-dimensional spacetime to 2-dimensional spacetime (imagine as a thin sheet of rubber) and now imagine normal sine waves (sine waves for simplicity) on this rubber sheet.

But wait, I mean, what creates these wiggles in the fabric of spacetime? The primary sources of these gravitational waves are actually emitted by events like the collision of two stars, white dwarfs, neutron stars, and even black holes. In the next chapter, we will discuss this in detail and it's one of my favorite chapters of this book. These Waves of gravity are huge enough, to not consider their effects on normal-day objects. But with the help of special equipment, we actually have detected these waves of gravity. The name of this equipment is LIGO. Which has L-shaped arms and we have detected many gravitational waves traveling throughout the universe by ligo. We even use Pulsar timing arrays (PTAs) to detect even larger gravitational waves, the origin of which is still unknown. For now (October 2023) we have detected only one measurement of these gravitational waves captured by PTA's. The gravitational waves can leak a lot of information from their source like colliding neutron stars and black holes or whatever that creates them. The gravitational wave created just after the birth of the universe can provide a lot of information about the past of our universe, confirming various theories and even totally changing our understanding of the laws of physics.

Most of the information we have gathered about our universe is by using one source alone, which is light! We know what our stars are made of, we detected exoplanets and photographed two supermassive black holes to test our theories, we discovered that our universe is expanding, etc. Again, gravitational waves like light also travel at the speed of causality, which is the maximum speed possible in our universe. The first-ever confirmation of gravitational waves at LIGO has created a new branch in Astronomy that is not based on measurements of light, "Gravitational waves astronomy"! This has broadened our horizons at a cosmic scale. The intriguing for this chapter is that one can build a telescope by gravity! What?! Yes! It is again discussed ahead in this book. In the next chapter, we will be discussing more about gravitational waves and this is going to change your view on how you looked at the gravitational waves in this chapter.

EIGHT

MASS TO ENERGY TRANSFORMATION WITHOUT NUCLEAR FUSION!

We all know or heard of energy obtained by nuclear fusion. The energy obtained in nuclear fusion is from the mass of the part of the main material used in fusion. A small part creates a chain reaction and creates a huge amount of energy. Well, this energy comes from nothing but mass. It is correct to say, that mass itself is an energy.

When we talk about energy in daily life and even in science, we all mean the ability to do some work. When we ask how much energy can we get from this fuel in Joules or any other available unit, we think of how much work can a system do with this energy input. So, energy is simply the ability of a system to do work. But did you know the mass which we measure in Kilograms is actually energy? I mean you can actually use a kilogram of anything, convert It into energy, and do some work. This energy created is enormous in amount. A kilogram of mass can generate 9×10^{16} (Nine followed by sixteen zeros!) Joules of energy! Let me tell you how much this energy is. An average house in the United States uses 3.88×10^{10} (388 followed by 8 zeros) Joules of energy every year. So now by dividing this much energy by the amount of energy we receive by converting the mass of 1 kg into energy, we get 2.3 million houses! So energy obtained just from the mass of 1 kg can power 2.3 million houses for an entire year!

This is how much energy a small quantity of mass stores. In nuclear fusion, only a small amount of mass of the entire nuclear fusion is converted into energy. We still don't have not built any technology in the world, which can convert any mass directly into energy. Nor, do our observations in the universe suggest any system doing this itself. But, we have seen a tremendous mass liberated in the form of energy via gravitational waves during the collision of supermassive black holes and neutron stars.

Let me ask the reader a question, when anything comes under the influence of gravity wave it gets stretched (provided we have a thing large enough to see the stretching effects). Now here the work of stretching has been done. But where did the energy to do this work come from? It's the source producing gravitational waves itself! Imagine masses of two colliding black holes thousands of light years away from Earth reaching the earth in the form of gravitational waves, stretching and squeezing it.

Let me give you an example of how much energy is liberated in the form of gravitational waves in the case of two colliding Black holes. Let's say we have two black holes on the verge of collision, with a mass of one being 36 solar masses and the other 29 solar masses (1 solar mass = mass of a sun). When these two black holes collide, they form one larger black hole with a mass of 62 solar masses. If you calculate the mass after collision, it should have been 65 solar masses. Then where did the rest of the 3 solar masses go? It got converted into pure energy! This energy is released in the form of gravitational waves. Remember, in an earlier example, we could power 2.3 million homes with energy obtained from a rest mass of just 1Kilogram. In this example, we have a mass of as many as 3 suns. The energy we obtained here is 3×10^{48} joules (3 followed by 48 zeros. This tremendous amount of energy!! Want to know in physical terms how much this energy is? Okay, this energy is more than the energy emitted (in the form of light) by all of the stars in the entire universe! Isn't this just so mind-blowing?

There are various methods to harness energy from black holes. All this although are theoretical. A very highly advanced civilization will use these techniques to carry out its massive energy needs. This process of harnessing energy from a black hole is again explained in detail in my book "Astrophysics for Non-mathematicians". Let's now go with how can gravity be used as a lens in the next chapter.

NINE

THE LENS MADE OUT OF GRAVITY!

The light always traces a straight path, that's what we all know. And there is no exception to this law. When light gets bent in any medium like water, it is still seen to be following a straight path in the scientific treatment. But what if the entire fabric of reality is getting curved? Then no matter what is connected to that fabric of reality, one has to follow that path, including light. The mass of the earth is enough massive to create a spacetime curvature (gravity) to make us feel the value of gravity the way we feel. But this spacetime curvature still doesn't bend the light.

The image below shows the Bending of light rays due to spacetime curvature.

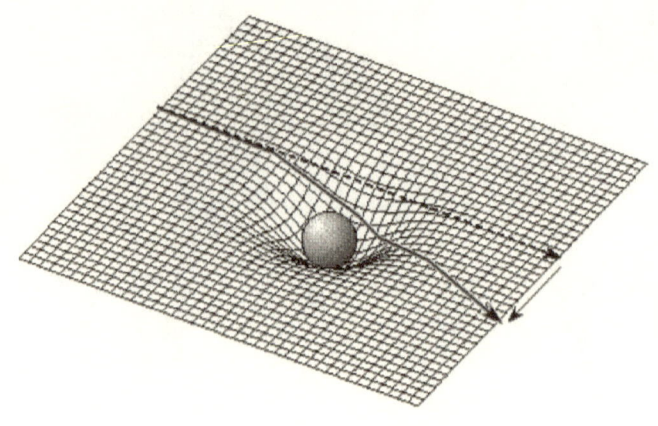

But for spacetime curvature like the one possessed by the sun, we have seen the apparent shift of position of the background stars during solar eclipse. During solar eclipse, because that's the best time to see background stars near to the sun. As majority of the light gets blocked by the moon during eclipses. So, in fact, light can be bent when it encounters something massive and dense enough to possess enough spacetime curvature to bend its path. So the gravity of the celestial body bending the light actually acts like a lens. This lens is what we call the "Lens of gravity" or simply the "Gravitational Lens".

This lens if used systematically, can act like a giant telescope for us. It can also reveal other celestial objects and galaxies that are beyond any other celestial body, and we are unable to see them. The gravitational lens also tricks us by showing the virtual image of a star that is far beyond its original position. For example

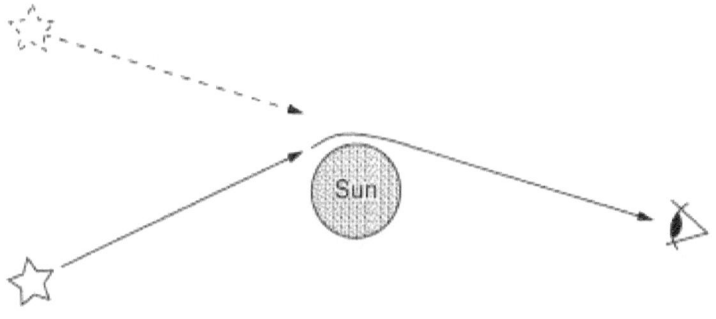

See below, the image captured by the James Webb Space Telescope. The distorted images of galaxies in the picture are due to the lensing effect of other celestial bodies or galaxies, coming in the path of light.

Black holes are black making them harder to find in the vastness of the universe. Thanks to gravitational lensing produced by these black holes, we have found so many black holes because of the lensing effect they have on their surroundings.

The image below shows the light from the galaxy lensed by some celestial object in its path in a ring shape. This ring is also called the Einstein Ring.

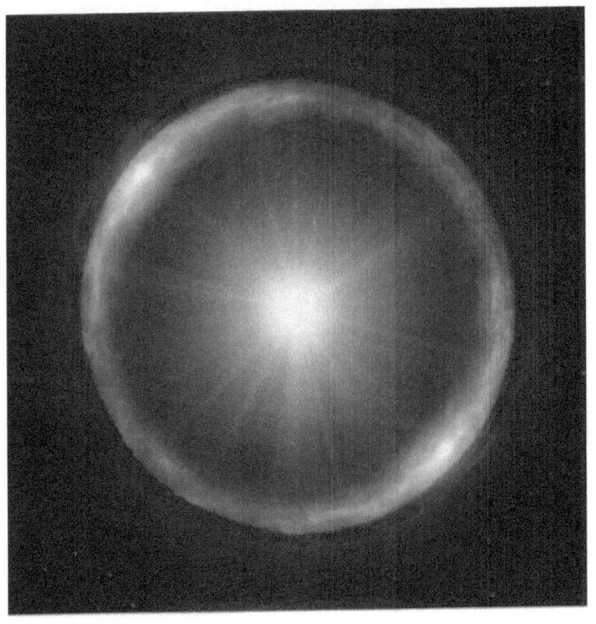

The things which I am discussing here in this book, many of these might not even be known to few readers to ever exist. There are still so many things, which are still a mystery in science. For example, there is a mysterious substance present all over the universe. And the surprising fact is that it's much larger in quantity as compared to Normal matter. As per our knowledge, we only know that this substance only interacts via gravity. So these substance is present in the universe more than the normal matter in quantity. But we are hardly able to detect it. We will talk about this substance in the next chapter.

TEN

THE MYSTERIOUS SUBSTANCE THAT IS MORE ABUNDANT THAN NORMAL MATTER!

There are four fundamental forces in this universe.

- The Force of Gravity
- The Force of Electromagnetism
- The Weak Force
- The Strong Force

Out of these forces, the strong force is the strongest, and the second strongest is the electromagnetic force. Now I guess, the reader has guessed the third one. But no! It's not the force of gravity! It's a weak force. And the weakest of all fundamental forces is the force of gravity

The matter we are familiar with, which we see around us, the same matter by which our body is made is in general terms what we call normal matter. We can see this type of matter, we can touch it because of the repulsion between the charge of atoms at the tip of our finger and the charge of atoms of the things we are touching. And this is how we feel the things, we don't touch matter, we just feel the repulsion between us and the thing we are touching. This repulsion in atoms is because of the force of electromagnetism. We can see the light of the stars in the night sky, again, this is based on the force of electromagnetism. We can feel gravity, and this is because of the force of gravity. When we taste something or smell something, it is again due to the force of electromagnetism. When we talk about the rest of the two forces, which are strong and weak forces. They have their main mechanism limited to the subatomic scale due to the limited range of their force fields. In fact, when matter makes sense to matter, it is via all or by one of these forces.

There is yet another special type of matter scientists speculate to exist. This type of matter as per our observation is only seen to be interacting gravitationally. This makes this matter invisible, since no electromagnetic interaction means no light. It does not interact strongly. It might interact weakly, as we have no good reason to contradict this. But all we are confirmed about is that it interacts gravitationally. Again a reason why we could spot it in the first place. We have spotted this matter, we just make an estimation about how much of this matter is present in what galaxy. And since it interacts gravitationally it must have mass-energy. This matter is what we call "The Dark Matter". Dark because we can't see it. Our universe contains 26% of Dark matter and just 5%

of ordinary matter. Wait! What about the rest of the 69%? Well! that would be discussed later in this book.

Provided we have so much dark matter in this universe, it can be anywhere anytime. Yes, but it will not make its presence feel at the local scale, since the only force it interacts with is gravity, which is the weakest force of the universe. Dark Matter is still a mystery to us and is yet not understood. But we know that there is something that exists within galaxies that makes them compact enough to form. This is what we hypothesize about Dark Matter. Let me drop a very very intriguing fact that will blow your mind. You probably laugh at me if I directly say it. But I am going to because it is the way it is. Okay, so that fact is that "We all are waves!". Wait what!? Yes! This will be discussed further in this book. For now, let's get to the new topic of this book. It's "Our universe". And you are going to learn everything new about it in this book. So let's move ahead and learn about our universe like never before.

ELEVEN

THE BIRTH OF TIME, THE BIRTH OF OUR UNIVERSE, AND THE BIG BANG!

The birth of the universe is the birth of time, the birth of time is the birth of the universe! Our universe is "spacetime". If time does not exist, the universe is nothing special than the surrounding hyperspace.

Okay, Did you ever wonder what is outside the universe? For simplicity let me assert that whatever exists outside the universe is hyperspace. Now what is Hyperspace? So when we study dimensions, a point is a zero dimension, a line is 1-dimension, A plane is a 2-dimension, the world around us, where everything has length, breadth, and height is 3-dimension or simply say for example a cube is 3-dimension, our universe is 4-dimensional although our mind can't think more than 3-dimensions. Anything more than 4 dimensions is what we collectively call "The Hyperspace". So how many dimensions are there in total? Infinite! So what outside our universe is the hyperspace?

In this chapter, we are going to discuss everything about our universe from a fifth or higher-dimensional perspective. The universe in which we reside appears 3-dimensional to us due to the constraints of our minds. But when we see the universe from extra dimensions, it becomes easier to understand it. To explain the birth of the universe in short and layman's terms, the texts below are oversimplified, mentioned in the form of a story for the reader to get a basic idea of the concept.

There exists the hyperspace of an infinite number of dimensions, the first 4 of these dimensions *decided to become something special. And all they did was make the 4th dimension flow and it became "time". And since all of these 4-dimensions were connected, all the dimensions started flowing with the fourth one, creating something called **"The Spacetime"**. Which is also called **"The Universe"** by the consciousness residing in it.*

So the birth of the universe is the birth of time! And the birth of time marks "The Big Bang". So, *as soon as the spacetime became special from the surrounding hyperspace, matter-energy started creating itself. Simultaneously while this happens, our universe goes into drastic expansion! After that, the accelerated expansion and the universe as per our observation is still expanding to date.*

The birth of time, the birth of the universe, and the Big Bang are the same thing. Initially, just after the big bang, the universe was very very hot, until it cooled down due to the expansion of the universe, forming the first stars, supermassive black holes, and galaxies. *But wait! Why did Spacetime decide to become special?*

Why our universe is expanding? And how did matter energy come in the first place just after the Big Bang? Why it ever happened? These might be our common questions as humans. We are going to discuss all of these soon in this book. And all we need is some new theory for it. It might sound like a movie but we need some quantum physics for it. Don't worry but, this is astrophysics for everyone! Sir Richard Feynman who is one of the greatest scientists and teachers of all time said "If you can't explain something in simple terms, you don't understand it". And I do my best to understand the things I study. So the quantum physics part of this book is going to be like fiction but factual rather than a boring physics lecture (For people who don't like maths) . So just bear with me for the next few chapters. I am doing all of these just because I want to give that cosmic perspective to everyone. Which teaches us about our significance in this universe as conscious beings and about life. So, before we move on with quantum physics, let's discuss two more topics about our universe I am curious to share about in this book.

TWELVE

HOW FAR IS THE EDGE OF THE UNIVERSE AND WHERE IS ITS CENTER?

I know it is hard to believe in something that can't be seen. We can't see time, but let's say we believe it exists. But then all we do is treat space and time separately, which is incorrect. Unless and until we visualize our universe correctly, we will keep asking wrong questions and find answers for them. Our universe is just not space, it's just not time, it's instead space+time or spacetime. The reason that we can't see time connected to space is because we treat spacetime as space and time as they are not connected. Let me explain to you the difference between these two notions here.

If I tell you to point something by your fingertip and ask what you are pointing at, then you would say it's a point in space or some location. But it is not just location but also an instant in time. Better to say its location or point in space and instant in time combined! This combination is called "An event". And this is what happens all over the universe. This might be difficult to understand, but this is how it works. Our universe is spacetime and all the points over it are the points of spacetime. And a point of spacetime represents an event. So let's say, for example, you have to catch your train, and now the question you ask is at what platform is the train coming at what time. And so what happens in reality is you catch a train on a particular platform at a particular instant in time. So catching a train is an event, which is a point in spacetime. I want the reader to think about this idea of treating our universe as spacetime and every point on it as an "event".

So our universe is the fabric of spacetime. When we ask where are the boundaries of the universe, we ask for a boundary after which no spacetime fabric exists, which again means no events exist! So instead of asking how far is the edge of the universe, one should ask "What event marks the end of the fabric where all the events happen?". Before the big bang, there was no spacetime. So the event of Big Bang marks the event where chronologically all events end. And if you ask, what marks the end of the fabric of events in the present time, it all happens at a singularity of a black hole. There is a theory that the universe is going to end just like it was created and it's going to collapse reversing its expansion, this is "The theory of Big Crunch". If this theory is true, the Big Crunch like Big Bang also marks an event but in the future where all the events ends. Again, this is just a theory and we have no idea about the fate of our universe scientifically speaking. If this is to happen, the very very advanced civilization may tackle this fate by traveling to some other universe (which will soon be discussed in this book).

WHERE IS THE CENTER OF OUR UNIVERSE?

Using the discussed notion of how to visualize the spacetime. The question we should ask here should be "What event is the center of our universe?". Looking from higher dimensions this event which is the center of our universe lies somewhere in between the Big Bang and Big Crunch (if the Big Crunch theory is true). Although we don't know what that event is.

In the next chapter, we begin with Quantum Physics for everyone.

THIRTEEN

IS REALITY REAL? THE BIZARRE WORLD OF QUANTUM PHYSICS.

Imagine a world, where matter particles can move through a wall or any barrier without crashing into it. Imagine a world where the same atom can be present at two or multiple points at the same time. This all starts with the question of whether we are waves or a particle. I mean this question might first sound stupid, but we have seen the particles behaving like a wave! We have conducted various experiments and have found that matter at some events behaves like waves and like a particle in others. Also, light which we have always been thinking about as a wave has been seen behaving like a particle in some cases, for example, the photoelectric effect. So, what matter really is, a wave or a particle? It is both! In some cases matter behaves like a wave in others it behaves like a particle. For example, if the matter doesn't behave like a wave, the radioactive decay of alpha particles in nuclear fission can't be explained. And there have been many examples like this.

The real world which we know, which we can see is the outcome of the decision of things at the quantum level to behave like a wave or particle. And this is how reality is created.

But what decides how things at the quantum scale (subatomic scale) should behave, like a wave or a particle? Or should we ask "who" decides how things at the quantum scale (subatomic scale) should behave, like a wave or a particle? I mean is that an intelligent conscious being like us? Read the below texts

Studies say if the value of a pie is different even at the 10th of its decimal place, the existence of intelligent conscious beings would never have been a possible case. In fact, if the value of any of the fundamental constants of the universe was different, the conditions are then not suitable for the existence of conscious beings. This universe is very very very finely tuned for the existence of conscious beings. Where every star, every planet, and every fact of the universe is in a way that supports the evolution and existence of intelligent conscious beings.

"If light decides to not behave like a particle during the photoelectric effect, no plant would be able to prepare its food from the sunlight. No plants means no life. So who decides what light should behave, like a wave or particle at any specific event?! The finely tuned universe!"

See the image below:

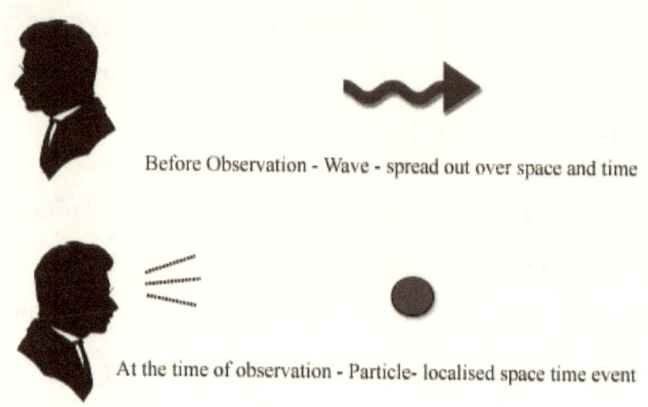

Before Observation - Wave - spread out over space and time

At the time of observation - Particle- localised space time event

Although this image exaggerates the process, the indirect meaning of this image is true. And you would agree to this at the end of the next chapter.

So have we as conscious beings made wave functions of various stuff behave like particles In some cases and as waves in others to make the universe the way it is? Let's discuss that in detail and understand it better leaving no traces of doubts and misconceptions in the next chapter.

FOURTEEN

WHY THIS UNIVERSE IS THE WAY IT IS?

So have we as conscious beings made wave functions of various stuffs behave like particles in some cases and as waves in others to make the universe as we see it? In a way that everything creates itself so to allow the existence of intelligent conscious observers in general. So in short, **The observers create the reality**. I have already reached a conclusion on this topic saying "The observers create the reality". But if I end this chapter here now, I am going to create misconceptions for many readers. So let me take you to the detail of what this observation exactly is.

IS THERE A MOON WHEN NOBODY LOOKS?

Moon is made of material particles, all of these particles at the subatomic scale showing quantum effects, sometimes being waves and other times being a particle. But when more of these quantum particles clump together or the bulkier they get, the less wave-like behavior they possess. This is the reason why your kitchen does not behave like a wave when you are in your bedroom. Well, if the moon started behaving like a wave, there are chances that you find the moon orbiting Saturn at one time and Mars at another or simultaneously both of them at the same instant. This phenomenon does not support the existence of the existence of the observer in the universe. Again, this phenomenon is against all the reasons this universe is created the way It is, and this is not how our universe works. Quantum mechanics becomes less of quantum mechanics when we talk about normal day objects like planets, stars, or anything that is not at a quantum scale. But since everything starts from the quantum world, it's a deciding and main factor the reality is the way it is.

So, observation makes this universe the way it is. But what is this observation in general? The observation that the evolution and existence of conscious observers are sustained in the universe! This is what the universe and everything from smallest to largest scale take care of. Or simply the observation creates it in that way.

So, when we ask the questions like *Why did Spacetime decide to become special? Why our universe is expanding? Why it ever happened?*

There is only one answer to this question. Look at the figure below :

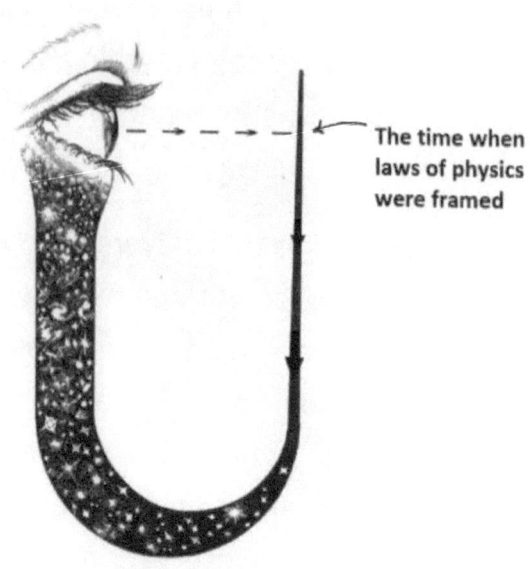

The time when laws of physics were framed

This concept is called the participatory universe, with participants being no one but "the observers"!

It from bit symbolizes the idea that every item of the physical world has at the bottom...an immaterial source and explanation and explanation....that all things physical are information-theoretic in origin and that this is a participatory universe

- John Archibald Wheeler

The answer is that the universe is just the way it is because it must create itself in a way that supports the existence of conscious observers like us. In a matter of a few chapters, we have totally changed our perspective to look at our universe. Well, this how astounding our universe is. The concept we discussed in this chapter is just a small and basic part of the relationship between an observer and the universe, in reality, it's a far more complex concept to think. And we still have very less understanding of it.

FIFTEEN

THE LAWS FOLLOWED BY EVERY SYSTEM PRESENT IN THE UNIVERSE.

Every system in the universe i.e. stars, solar system, planets, body system, systems like black holes, etc. All follow a set of 4 laws (at least in our universe). These are called the "Law of thermodynamics". Let's go with every one of them shortly.

THE ZEROTH LAW OF THERMODYNAMICS: When three things of different temperatures come into contact with each other, the three as a system tend to have a common, same temperature. As easy as it may sound.

THE FIRST LAW OF THERMODYNAMICS: The first law of thermodynamics simply says that the mass-energy created in the Big Bang (The birth of our universe) is the only mass-energy available in our universe. This amount can't be increased or decreased. It always remains constant.

THE SECOND LAW OF THERMODYNAMICS:

The information backed with some law of physics created by a system per unit of time always increases in the universe (No worries, you can skip the explanation and move to examples given below for simplicity)

If you couldn't grasp the above explanation, just refer to the examples below:

The second law of thermodynamics says that the engine making no sound is not possible in the universe, It says that a perpetual machine is not possible (a machine that drives itself forever is not possible). For example, imagine a water wheel, with a few buckets attached to it, with two reservoirs, one up and the other down. Every time the filled bucket turns the wheel, a new bucket gets filled up from the below reservoir, and the rest is deployed in the upper reservoir. This creates a perpetual water wheel. But again this is not possible because every time the bucket scoops up new water, it crashes into water, producing friction that destroys the perpetuity of this machine. Again, the second law of thermodynamics proves itself true.

The second law of thermodynamics is again the reason why everything in the universe ages.

But wait! Whenever we launch a satellite into space, it doesn't need fuel to maintain its velocity once it enters space, far from any celestial body. So do we again create a perpetual machine here? No, friction again takes over here. Friction in empty space? Because empty space is not empty, at a quantum scale there a particles called virtual particles that always produce some friction even in the empty space!

The third (last) law of Thermodynamics is absolute zero temperature is not possible. Even in empty space 0 temperature is not possible. So where does temperature come from in empty space? From nowhere but from the quantum scales where virtual particles themselves contribute as a temperature. This temperature due to virtual particles is also called the zero-point temperature.

Let me tell you an intriguing fact, we all know that our universe expands, but did you ever know that it does expand faster than the speed of causality? So does it break physics?! The answer is no, let's know why in the next chapter.

SIXTEEN

UNIVERSE BREAKING ITS OWN LAW OF PHYSICS!

No matter what, nothing in the universe can break the very fundamental law of physics. For example, the fundamental speed limit of the universe. The speed of light. Nothing in this universe can travel more than this speed. Except for one, the universe itself!

Yes, our universe has been expanding since the Big Bang, and at some time in the epoch, it started showing accelerated expansion, this accelerated expansion of the universe gives us an energy anomaly in physics. Since the total energy in the universe always remains constant in the universe. To cover up this anomaly as a scientific mystery, we have named it dark energy like dark matter this energy composes about 69% energy content of our universe, the next 26% goes to dark matter, and the next 5% to normal matter.

What again surprising about the expansion of the universe is that it is expanding so fast, to move galaxies far away from us more than the speed of causality. So is the universe really breaking its own law of physics? The answer is no. Because the galaxies that appear to move away from us are not actually traveling more than the speed of light. Okay, let me explain it to you in brief.

Firstly, to get out of the misconception, when we say the universe is expanding, it's not that the universe is stretching itself, instead the new fabric of spacetime is getting created at every single point in the universe. This creation of spacetime can't separate Earth from the sun and neither the sun from the Milky Way as it creates itself. Its effects are only apparent at many light years away from any reference frame.

Coming to the main point, when anything travels in this universe with some speed, the spacetime locally starts changing for him. In the case of expansion of the universe, the galaxies move farther away from us at the speed of light, but they are not traveling as their spacetime doesn't change due to this movement. These galaxies are just residing away from us due to the creation of new spacetime in between. So that's why nothing in this universe has and can ever travel more than the speed of causality. But we now have a part of the universe residing far away from another more than the speed of light. So if not the constituents of the universe, is universe itself violating its laws of physics? The answer is yes! But its all okay if universe violates the law of physics. So the laws of physics can't be violated by something that exists within the universe, but the universe itself can do that. I know we have reached the conclusion for this chapter, but there is one thing I am curious to share about our universe. We all have seen the map of Earth, but have you ever wondered what the map of the universe would look like? Yes! The universe has its map! And this is how it looks :

The image displays the universe as it was just 3,80,000 years after its birth. The color difference shown in the map is the temperature difference, This map is what we receive from all around the universe in form of microwave spectrum.

SEVENTEEN

HOW DOES A GRAVITY TRACTOR MOVE AN ASTEROID FROM ITS ORBIT?

Let's say an asteroid larger than a football ground is heading toward us, so is that the end of the world?

Well, that depends on the factors of what the asteroid is made of, its angle of collision, the mass of the asteroid, and the velocity of the asteroid. If the asteroid is made of various metals then it is more hazardous than the ones which are mostly composed of carbon. And if the collision is taking place say at an angle that is less than 90 degrees then the kinetic energy (energy possessed by every moving object) would be wasted in peeling off the surface of the earth and less violent shockwaves would be created and when the angle of collision is 90 degrees then the most of the kinetic energy of the asteroid is transferred in the form of a shock wave and rest in the form of heat when it penetrates the earth surface. When these shock waves make their way to the oceans then tsunamis will get triggered (flushing out a whole city or country or even the entire world). I should describe the role of mass and velocity collectively as the momentum (= Mass x velocity) is what plays the role here. When the asteroid hits the earth its momentum gets transferred to the earth. I guess an impulse (momentum imparted in a very short interval of time) would be a great replacement for term momentum over here. The entire humanity is in danger. The threat of such asteroids always remains (Although less probable, can or will happen).

How do we detect the asteroids?

Look up in the clear night sky until you watch any falling star and that is how we do it! These falling stars are just meteorites which are just the chunks of the larger asteroid or any comet. Asteroids really don't have their own light so unless the rays of the sun fall over the asteroid and some traces out of it get reflected, we are not able to target them. And to check for its size, shape, and orbit we use radar technology. Once the trajectory is calculated, one can predict the future path of an asteroid.

How can we shield ourselves from these hell stones (real ones! literally!)?

Asteroids may move at a velocity as high as 30 km/s and even more! They have violent trajectories and irregular surfaces. These factors make asteroids harder to detect. And the best moment to tackle any asteroid is only when it's more than 7.5 million kilometers away from us otherwise they are considered potentially hazardous ones (provided they have enough mass and velocity to create massive destruction) which are even harder to deal with if we aim to deflect it. But how to deflect an asteroid off the track? This how :

By Nuking it:

Yes in a situation when an asteroid (potentially hazardous one) is calculated to hit the earth, we can nuke it to either move it to the safer track or to destroy it. It depends on the mass of the asteroid to conclude how many nukes should be detonated to move it off the track (according to Newton's third law). But still, the problem is that even if one turns an asteroid into pieces by nuking it, the meteorite or micrometeorite formed due to the destruction of the asteroid may destroy our satellites. Thus there is a huge loss in the economy. But still, human life is important. It's an effective way to destroy or boost the asteroid off the track.

Deflecting an asteroid by the light :

If I tell you that you can deflect a metal sheet or any object by just making some light incident over it, then probably you may say I am kidding! But no! I am not. Light imparts momentum like a normal traveling object. But light has no mass! But it has energy and as energy is the same as mass, light has its own momentum to impart. The pressure generated by the force imparted by the light due to its transfer of momentum is called the Radiational Pressure. An asteroid can be moved to some other track in this way. It's a good way of deflection as all we need is only high-energy light rays to impart great momentum over an asteroid to deflect its path.

Deflecting an asteroid by a gravity tractor :

Two masses always have the property to attract each other by the means of gravity. Let's imagine an asteroid (potentially hazardous one) moving towards the earth and we try to deflect it by means of some mass by having a gravitational tug between that particular mass and an asteroid. This mass is a moving spacecraft with an ionic propellor sent by us which is set to hover near the asteroid. The spacecraft is placed in such a way that its gravitational pull on the asteroid is perpendicular to the plane of the asteroid belt. If the spacecraft and an asteroid if considered as a system then a slight change in spacecraft velocity will affect the spacecraft and asteroid system's momentum as a whole, moving it off the asteroid orbit plane. The force due to gravity due to spacecraft may be very low on the asteroid but over the years the consistent pull can drag it from its orbital plane thus securing the existence of our planet Earth or maybe of humanity as a whole.

EIGHTEEN

TIME TRAVEL AND THE SPACETIME INSIDE A BLACK HOLE.

In the first few chapters of this book, I explained what time exactly is, It's just a "flowing spatial dimension". So as we move in space, can we move in time? I mean, can we go into the past and change the present or future? This is what we or many know as time travel. Well, the answer is yes if you time travel in the future and no if you are going in the past. Let's discuss every question regarding this.

There is one thing by which "time" differs from "space" that it always flows carrying "space" along with it. This flow is unidirectional and always moves in a specified direction in Hyperspace. This nature of time to flow is called the "timelike nature". And the ability to move in 3-dimensional space is called "spacelike nature". Let's learn ahead, of how can we travel in time in the future and not in the past.

HOW TO TIME TRAVEL IN THE FUTURE?

So, to explain in simple layman's terms, when anything makes time tend to become spacelike, the space for him tends to become timelike (one starts to move in space). What I discussed here is what a black hole does. So that's why time slows down when one goes near to a black hole. And when one goes far away from the black hole, after nearing it, he will discover himself to have time traveled in the future. This happens because from the higher dimensional perspective, when your time tends to become spacelike, the flow of time slows down for you, although the time of the universe far away from the black hole always ticks at a constant rate. And when one returns from his voyage around a black hole, one discovers, that so much of time has already passed in the universe while only a little of his. Technically he has time-travelled into the future by doing this. The reason behind this is that when time tends to become spacelike, one technically as the end result gives himself the freedom to move in the future like we freely move in space. The reason behind, this is that when time tends to become spacelike, one technically as the result gives himself the freedom to move in the future like we freely move in space. But still one doesn't have the ability to move in the past as we are going to discuss in this chapter ahead.

But every time we can't manage a black hole, so what can we do as an alternative to this? Simply just reverse the order of the process :

"When one starts to move in space, space tends to become timelike and time tends to become spacelike"

And that's why as we go near to the speed of light, time slows down for us and when we stop moving and come back to rest, we discover that we have travelled into the future.

I had used the phrase "tend to" in the texts mentioned here. That means space is becoming time-like but has not yet. And time is becoming spacelike but has not yet! Because when it does that, the "spacetime" ceases to exist! And that's why we see nothing beyond the event horizon of a black hole since spacetime simply ceases to exist there.

WHY IS TIME TRAVEL IN THE PAST NOT POSSIBLE?

When time tends to become spacelike, the time in spacetime still moves in the forward direction (past to future). So, until you are a part of spacetime (our universe), no matter how spacelike your time becomes at a "certain level", time always be moving in one direction (past to future). This certain level is marked by the event horizon of a black hole and the speed of causality or the speed of light in our universe. The physics of this is what we are going to discuss in the next chapter. So all one can do in the spacetime universe which we belong to, is slow down time for us, although time is always going to move in a forward direction in the case of our universe and for anything that exists in our universe. So during time travel in the future as discussed above, when we near a black hole or move near to the speed of causality, time slows down for the traveler while maintaining its arrow. Traveling in past means changing the arrow of time by 180 degrees, which is not possible in this universe. In the next chapter, we will be discussing what happens when time becomes spacelike and space becomes timelike and how this would look like to a traveling observer. The next chapter is super interesting. So just bear with me as we near the end of this book.

NINETEEN

THE STRANGE WORLD WITNESSED BY THE LIGHT AND GRAVITONS.

As we all know, time does not elapse for light since it travels at the speed of causality. *Whenever anything travels at a speed of causality space becomes timelike and stays spacelike simultaneously! And time becomes spacelike and stays timelike simultaneously! This is what even happens just the rim of the event horizon of a black hole.* Now, let me give you an idea of what this world looks like.

The light from the sun takes around 8 minutes to reach us, but if you ask the photon about the total time elapsed since it was emitted from the sun, the answer would be "no time"! So for light, anyone who looks at the sun waiting for the light to arrive for 8 minutes, the person in light's perspective was always predestined at that moment at that spot to see light! I want the reader to think about this. In our universe, the future state of the universe is represented as a wave function instead of a discrete predestined outcome. So what physics is going on here? The Light is now in a very exceptional state of the universe, it's the state where time has become spacelike as well as stayed timelike at the same time showing the dual nature. This makes the existence of light stretch its way throughout all future spacetime until light exists in the spacetime part. This stretch only happens from the time light has created itself and towards the future, since light still has a part of it in the real universe. So unless and until light exists in this state, it will mark a predestined existence throughout spacetime, not leaving its fate to collapse of the wavefunction. Not only light but this same happens even with the gravitons as they travel at a speed of causality. This happens with anything that sits on the rim of the event horizon of a black hole. In the case of the event horizon, there is still a further story to tell. After the event horizon, space actually becomes timelike and time actually becomes spacelike creating a spacetime to timespace swap! Let's discuss more about this in the next chapter.

TWENTY

AFTER THE EVENT HORIZON OF A BLACK HOLE.

In the preceding chapter, we discussed what happens at the Event horizon of a black hole. But what after that? At the event horizon of a black hole, time becomes spacelike and simultaneously stays timelike, it shows dual nature there and space becomes timelike and again stays spacelike showing the dual nature. But what happens when spacetime is forced further inside a Black hole, so that space becomes totally timelike and time becomes totally spacelike? Nothing, but end up creating what we want! We totally switch the spacetime of our universe to the "timespace" swap! In such a universe (timespace) the amount of time flown displays itself as a three-dimensional reality. Since time now has become spacelike, one can actually see in time (only past) as we do in space. Although one can see photons which have stretched themselves in the future. So, time has now become a 3-dimensional space. This is the same as space in our universe, except the fact that the past might reveal itself in some parts of this 3-dimensional timespace reality. And since now space has become timelike, the space becomes one dimensional time equivalent of the timespace swap. This timespace formed would just work like spacetime. However, this timespace universe after a black hole is very short-lived, as the spacetime gets shattered at the singularity of a black hole. But wait, one might have a universe entirely made of timespace?! But still, It makes no sense to differentiate it from spacetime, since now technically there is no difference between spacetime and timespace. Wait, what about a universe where there are 2 dimensions of space and one dimension of time?! Alright, let's discuss that in the next chapter.

TWENTY-ONE

DOES GOD EXIST?

When we talk about science, it doesn't actually deal with subjects like these. Since science only has a limited theory in its domain. There is much philosophy on the existence of god as well.

"Earth's gravity is just so rightly strong to create a soup which we call air to allow birds to form the way they are. And again making our skies enough blue for oceans to reflect the right amount of sunlight that will make the temperature so right for life to dwell and evolve."

But science when combined with philosophy, hints to us a lot about god. For example, quantum physics, which has given rise to reality itself, provides a solid ground to so many philosophical explanations about god. (For example, one can read the Schrodinger cat experiment). In fact, quantum physics seems to strongly emphasize the universal consciousness, which we call "God". Again as we read these lines before:

"If light decides to not behave like a particle during the photoelectric effect, no plant would be able to prepare its food from the sunlight. No plants means no life. So who decides what light should behave, like a wave or a particle at any specific event?! The finely tuned universe!"

So as a physicist and philosopher, one can see that physics needs universal consciousness to explain the universe.

In the next chapter let's see whether there is any life like us on some other planet in this universe.

TWENTY-TWO

ARE WE ALONE IN THIS UNIVERSE?

To this date, there is no solid evidence (Although many claim there are) of the existence of aliens. But still, is there any one of them out there? Any planet in another system in the habitable zone (a zone suitable for the creation of life) gives us the chance of having extraterrestrial life. Our planet Earth, which is in the solar system, is in the Milky Way which has more than 3200 star systems in it and still, there are more than three trillion estimated galaxies in the observable universe. Thus in that sense, it's highly probable for the alien lives to exist. Can Extraterrestrial life if exist be more intelligent than us? well! we don't know. Aliens may even be type 2 civilizations at the level of which technology is enough advanced to harness the complete power of any star. Aliens also may be advanced human beings. If they meet us we would have a great opportunity to learn from them. Also, they can be bacteria or any unicellular animal too.

Jupiter's moon Europa is quite likely to sustain life in it as there remains the liquid water underneath its ice sheets which have widths of several kilometers wide. Europa gets sunlight but not that bright to maintain the water in a liquid state beneath So from where does Europa get that energy to maintain the liquid water beneath the ice sheets? the answer is the tidal gravity of Jupiter. The tidal gravity or tidal force of Jupiter brings out some deformation in the shape of Europa, this mechanical energy is then converted into heat energy by breaking various molecular bonds providing some heat energy source to maintain the liquid water under the ice sheets of Europa. One also can compare this with the tennis ball when it gets hit by a racket. When a tennis ball is hit by a tennis racket, it gets deformed and in fact gets heated due to the mechanical deformation acted on it. In the same way, Europa receives its heat by which it maintains liquid water beneath the ice sheets on it. But here what is deforming Europa is the tidal gravity of Jupiter. This deformation is not as great as in the tennis ball analogy but quite enough to provide energy to the planet to keep water in the liquid form beneath the ice sheets on it. A mission is planned by NASA till 2024 to study in detail to investigate the suitable conditions for the existence of life on Europa and if we get some positive results there, then it would truly be one of our greatest achievements of humanity.

Well! what about the aliens from some other universe? Great idea right? If the multiverse exists, then we have a greater chance of life harboring in each of the individual universes in it. So if we are alone in 'our universe ' then there might be aliens existing in another universe too. What about the bulk beings? (Multidimensional beings) There is no proof of the existence of bulk beings. If God exists in 10 or 11 dimensions we may call him a bulk being or bulk aliens or whatever you like. There is a famous equation of aliens that gives the probability of the existence of extraterrestrial life in our universe. It considers every parameter like the number of habitable planets, the fraction of stars orbited by planets, etc. This equation is also known as the Drake Equation.

$$N = R_{*} \cdot f_P \cdot n_e \cdot f_l \cdot f_i \cdot f_c \cdot L$$

N = number of civilizations with which humans could communicate
R_{*} = mean rate of star formation
f_P = fraction of stars that have planets
n_e = mean number of planets that could support life per star with planets
f_l = fraction of life-supporting planets that develop life
f_i = fraction of planets with life where life develops intelligence
f_c = fraction of intelligent civilizations that develop communication
L = mean length of time that civilizations can communicate

The presence of us in this universe as the only life in existence in actuality is a rare case. If we are in reality all alone in our universe, could it be interpreted as there being some relation between the existence of humanity and the existence of the universe? Roughly speaking it could be! Even some observations of cosmic microwave background like the axis of evil suggests that we humans have some special role in this universe! But still, this is speculation and there is no solid basis proving it strongly. But until we investigate each and every suspected habitable zone for life, our hunger for searching for extraterrestrial life will persist.

TWENTY-THREE

THE FUTURE OF THE OBSERVERS CREATING THIS UNIVERSE.

Since mankind or conscious observers have been born, we have made tremendous progress in terms of technology. The more advanced we become in terms of technology, the more is our tendency to use more energy. It all started with burning trenches of trees. Now we are using fossil fuel, nuclear energy, and solar energy as fuel. Within the next 250 years, all the fossil fuels would be replenished.

Might be then, we would rely on the solar energy we receive and use water as our fuel. Yes, water can be used as fuel. By electrolysis hydrogen and oxygen in the water are separated. The energy for the electrolysis comes from the solar energy. Once the hydrogen and oxygen of water are separated they are brought back again to react with each other, this process releases energy and the exhaust that gets out is pure water. This is a clean fuel. All of these are expected to come within the next 50 years. But what about the time scales like the next 1 lakh years or even more, what about 1 billion years !? How advanced would humans be till then? This chapter is all about it.

The progress at a civilization level is measured purely by our ability to harness fuel. Major the source we cover, the higher we go in terms of technology and we see an increased energy need. Our sun is extinguishing its fuel soon and will lose all of its nuclear fuel, way later than that sun will swallow the earth in its expanding atmosphere. Then the option we will choose is to move to some other star to harness its energy. But what after we harness all the energy from all the stars in our galaxy? We will then simply move to another galaxy. Let me tell you in short, we will burn out every fuel in the universe. And in the end, what we are left with is just black holes. After harnessing energy from these black holes, when there is no fuel available in the universe, we might change our universe itself.

The one thing we say is for sure is that the best to best technology we ever built in the future is available right now somewhere in the universe. A star is a nuclear reactor, hawking radiation extracts energy from black holes, Galaxies reside away from each other using fuel of Dark energy, etc. All the things special we are going to do are that we are going to understand nature and try to replicate it for the benefit of mankind. In conclusion, we just have barely begun to explore this universe!

www.ingramcontent.com/pod-product-compliance
Lightning Source LLC
Chambersburg PA
CBHW031427210526
45464CB00005B/2079